DIRTFISHING FOR DUMMY'S

By: *William Roseberry*

Contents

Dedication

This self help guide is dedicated to all treasure seekers before me. **Thank you for your technology and knowhow.** You guys have blazed the trails, for me and many others. **Furthermore giving us a life fulfilled with fun and adventure.**

Introduction

In the beginning of writing this book my wife asked me, "Why metal detecting? I replied, "It's simple, ego". If anyone asks me, "Do you know anything about metal detecting?" I can then say, "I wrote the book on metal detecting". Then I will pleasantly surprise them with a free copy, and "say would you like to donate to my cause".

The First Metal Detector

This is the greatest story about the invention of the first metal detector. Not long after Alexander Graham Bell invented the first metal detector. It was then called the balance induction machine. President Garfield was dying from an assassin's bullet. The bullet was lodged someplace deep within his body. The bullet couldn't be found by any of the best doctors of that time.

This could possibly be the most important relic hunt ever by using a metal detector. All the doctors then were dumb founded by the ability of a simple loop of wires that could be rubbed across the president's body to find the exact location of a bullet.

Gosh, Mr. Bell must have sounded like a supper idiot, when he asked to perform a scientific experiment on our president, with his brand new balance induction machine. Furthermore he goes on to saving President Garfield's life with the first metal detector.

A Brief History

Ever since the invention of the metal detector man has had the ability, to easily find relics and treasure buried long ago by our ancestors. Before the invention of it, I was a very complicated and labor intensive process. We had to hand dig and sift through all of the dirt and soil. This was then the standard way to relic and treasure hunt.

Now the metal detector in all of its forms has become a staple tool used in archeology and law enforcement forensics. We now have the ability to quickly cherry pick through thousands of yards of debris and, discriminate all iron and trash from precious metals like coins and jewelry. The metal detector makes it a fun and exciting way to search.

The law enforcement communities now have the ability to, easily locate fired bullets and spent shell cartridges. They can also quickly find things like guns and knives that have been thrown into waterways. The metal detector in all of its forms has given mankind the ability to do endless tasks.

Be Responsible

To start this book out correctly, I must talk a bit about responsibilities. **It is the responsibility of all detectorist *to follow state and federal laws that govern the hobby*.** Not only the written laws but also, all of the common curtsey's of mankind in general. Respect thy neighbor, as thy self.

Theory of Operation

The metal detector is a pretty simple device used to electronically search for metals, buried under ground. To understand how this device works. You need to have a basic knowledge of electricity, conductivity and its resistance to flow. This information may be somewhat boring; **however it will get more interesting and fun as we go.**

The loop or coil is defined by my brain as two simple windings of wire, usually made of copper or aluminum. Each winding is coiled in opposite directions of one another. They are located at the end of a metal detecting device. The two coils are called transmit and receive. *Together they make up the loop*. Pretty simple huh!

The Loops Job

An electric signal is sent from the main oscillating crystal located in the main box. To the transmit coil. Then it is picked up through induction, by the receive coil and relayed to the micro processing chip. **Hold up a second this is just so complicated your now saying**. But it really isn't. Let us all play pretend and substitute a micro-processing chip for a simple light switch. That light switch is now wired to a simple noise making device, like a car horn or speaker. Every time the light switch is energized the horn or speaker beeps.

Our frequency is now equal to noise. The noise is sent from transmit to receive coil. The transmit coil is now singing loudly. But the receive coil is sleeping and can't hear the music. Why is this? Well remember the windings in the loop are coiled in opposite directions. This intern causes them to have a perfect balance. **The perfect balance causes silence in your headphones.**

Time to throw in any conductive metal like, **silver** or **gold**. Only then causes an electromagnetic field to become complete and active. This in turn causes interference and screws up the perfect balance in the loop. The receive coil now quickly awakens. It can now hear the transmit coils music. In other words, when the coil of the metal detector passes over a metal object the beep can be heard in the headphones. This is so simple guys. **Are you still with me?**

Frequency

Frequency is nothing more than an electrical signal that is sent in different wave lengths or pulse patterns. The timing of these on and off patterns is measured in hertz. The faster the pattern equals; *higher frequency*. The slower the pattern is: *the lower the frequency*.

The reason we use different frequencies with metal detectors is because some frequencies are more sensitive to different metals. Lower frequency is best when searching for higher conductive metals like silver and copper. Higher frequency is more sensitive to smaller pieces of metal that are on the lower conductive metal scales, like small gold.

Conductivity

Unfortunately the scales of conductivity change from one detector to another. Though they do all have the same basic circuitry. The scales differ do to different engineering designs. All metals are conductive to heat and electricity, but they all have a different conductivity signature. They all will transmit heat and energy but, their properties are different from one another.

This is where the metal detector takes advantage of the different conductive signatures. Inside the main unit of the detector there is a memory chip. This chip memorizes the metals different conductive properties. This allows you to choose what metals, you would like to find. **This is called discrimination**. *Discrimination is used to eliminate the trash targets.* Now let's put together a very generic scale of the metals conductivity signatures

Here is a conductivity scale of metal from highest, to lowest. **Let's first debunk a theory that gold is the most conductive metal.** I know you have all heard it. **Fact**: *It is not.* We only use gold in modern electronics today, because it has a great ability to not corrode or deteriorate.

Generic conductive metal scale

From most conductive to least conductive

1. ...Silver, unalloyed
2. ...Copper, pure
3. …Gold, unalloyed
4. ...Aluminum, unalloyed
5. …Zink, unalloyed
6. …Nickel, pure
7. … Brass
8. … Bronze
9. … Iron, pure
10. Carbonized Steele
11. Lead, pure
12. Stainless Steele

How to select your first metal detector

Ok, we now get to discuss the most confusing part of metal detecting. This is where everyone struggles to get good information. They struggle for good reason. People in general do not want to waste their hard earned cash on a metal detector that gives them a bad first experience.

This is what really makes or breaks the hobby. When newbies get a bad taste for detecting, they simply just give up. They feel that they have wasted money on a detector and, this hobby just isn't for them. Now, on the other hand, when they use the right machine, they will have a great first experience. They might get the **dirt-fishing** bug, like me.

So this is why getting good information is vital. Your first machine will be the most important machine you ever purchase. It will either get you addicted to the hobby. Or it will make you absolutely hate metal detecting. So make this purchase a smart purchase. A purchase that makes you happy, that you dropped some cash, on your machine. **Now if you do not take any information away from reading this book.** *Take this, because I have studied many machines and all different technologies.* I'm here to testify with honesty. That you do not need a thousand dollar machine to find gold rings, silver coins and, civil war relics. **This can all be found with a one hundred dollar entry level machine**.

With that said. If you are going to be prospecting for gold in its natural form, you will need a seven hundred to fifteen hundred dollar machine. This is for good reason. Because gold is usually found in its natural form combined with many other very conductive elements like natural iron.

Before starting with the good bad and ugly of major name brands, we must discuss research. I researched six months or more before purchasing my first machine. Did it help? Not much. Why? Because there is, just too much biased information published on the internet. **Everyone is trying to over exaggerate there machine.** *They simply do it to make a sale or buck.*

So take what you read with a grain of salt. Just ask yourself, is there a reason this article or blog is all about Garrett and doesn't discuss any flaws. Well I'm here to tell you that yes Garrett does make a good machine, but which one? They have over twenty models listed for sale.

This is my theory. *Go with the government.* Go with what the police departments use. You know that they have done extensive research on machines they have purchased. Ask yourself, what machines they use to find crime scene evidence. Like fired bullets, spent brass or weapons and what knots. You know government entities have done their homework. This is easy, in Rome do as the police do. (Little humor)

You will find as you get into metal detecting. **That simple is better.** *Fewer bells and whistles equal more time hunting.* Less time will be spent reading Chinese written phone book style dry product manuals.

These are some deciding factors when considering purchasing a machine. Things to consider defer from person to person. First let me just say Metal detectors are like guns. You would not use a twenty two to hunt an elephant. It is just like you wouldn't, use a fifty caliber, to hunt a squirrel. It's that plain and simple. A different detector is needed for a different job. We are very far technologically away from making one detector a use all machine.

Name brands verses cookie cutter machines.

Let's start with the mother of the modern metal detector. Fisher Labs. Now If Fisher labs are the mother. Who is the father? Did you say Sir Alexander Graham Bell? If so that's great. If not, please reread chapter one. So now you're thinking, wow what does fisher have to do with this? They do not even exist as an individual company. Fisher labs are very significant because Dr. Fiscar yea spelled much different than fisher. Is very important because. He designed the first portable metal detector for sale to the masses. So that is why they are important in the history of garage start up.

Think about it. They were able to design and reproduce circuitry for portable detectors. Then make them available for sale to the mass market. *We will discuss Fisher a little more latter.*

Features and functions

Features and functions can get a bit complicated. So we will only talk about the most important functions you will want to have on your machine. These functions and features are the most basic needed for a great first experience.

A **water proof coil** is a very important feature to have. There is water everywhere. Even if: It is only the dew on the grass. Don't think for a second that your entire detector is water proof. It is not water proof it may be water resistant. That is different from water proof. Water proof means that your detector can be completely submerged. Water resistance means that a rain drop may get on your detector without hurting it.

Discrimination is needed so that you do not spend most of your time digging trash. You can devote more time searching for precious metals. Or items that you want to find. All machines have different types of discrimination. Some have computer programs for different types of hunting. Like coin and jewelry, or relic hunting. Some have only iron discrimination this is good for prospecting. **Remember you do not want to dig, foil and pop cans all day**. *That will make for a very bad experience.*

Manual ground balance is another important feature to have. It is necessary to compensate for changing ground mineralization's and conditions. It is greatly needed for hunting black sands.

The pinpoint function is another feature needed. It is needed because it is important to find the very exact location of your target. So that you are not digging a huge hole to find a small target like a coin. You need to find your target without digging a huge crater in your local park or your neighbor's yard. That would be the quickest way to get ran out. Or even put in jail for criminal property damage.

The pinpoint feature allows users the ability to switch the detector from motion mode, to none motion needed to detect. Therefore the louder the scream of the detector the closer to the center of your coil you're the item is. Now the nerd talks technical stuff. The pinpoint button switches your vlf detector into bfo detector. That simple it's like having two detectors in one machine. The bfo detector does not need any motion to detect the object.

Your needs verses your wants is an important decision. . Sure there are a lot of bells and whistles available. There are so many options that I could not explain in an entire book if I wanted to. So this is where your budget comes into play. The limits on technology is only limited to your budget. But be advised. The limit on your finds is only dictated by where you put your coil period. Bottom line is money is not going to give you the advantage over your competitor. It is the hard work and historical research that will. In the end: After about six hundred dollars. You are basically only paying for hand warmers and paint. Unless: You are buying a high end pulse induction machine, used for prospecting. Prospecting gold in its natural form and element takes a more expensive machine.

Top secret detector sold by radio shack in the 90s

Now it is time to give you some valuable information, about your fist choice. I own two machines that are worth more than a thousand dollars. But I also own a machine that cost only one hundred dollars. This machine was sold by radio shack in the nineties. This machine was an entry level machine exclusively made for radio shack to sell. This detector has the worst name brand name of all U.S.A. made machines. You probably guessed it. The Bounty Hunter machine: Believe me this machine is a sleeper that hardly anyone knows about. It will blow the ace series detector line made by Garrett out the water by no comparison. It even has manual ground balance.

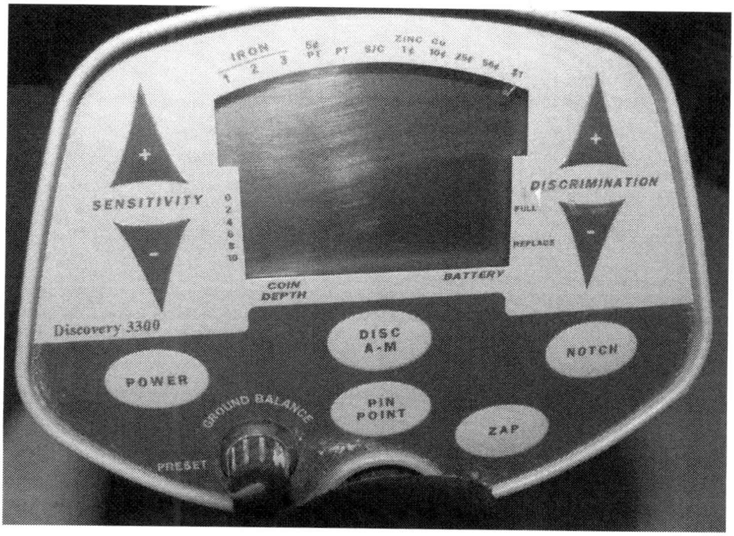

One would never suspect a Bounty Hunter machine to hang with the so called Big Boy's. Seriously now here is the secret. Fisher labs were purchased by first Texas Company that makes bounty hunter. Now during this transition period Fisher lab engineers designed The Bounty Hunter Discovery line. I do believe it was strictly made for radio shack. At the time it was just listed at an over inflated price. I think it was about three hundred and fifty dollars. It was soon clearance out with hardly any sales. The exact model I'm referring to is the Discovery Thirty three hundred series made by bounty hunter.

Now that the cats out of the bag. Let me just say that this detector is a beast. I bought mine from a local pawn shop for eighty dollars. I see them online every once in a while for a hundred dollars or less. Defiantly buy one of these detectors if you get the chance, especially for a first machine.

Please do not confuse this information with other Bounty Hunter machines. The only machines that they do make that can compete, is the platinum and above, though they are grossly overpriced. Plus they are much more complicated to operate... **Keep it simple and cheap and you will have a great first experience.** After that first great experience you will not have a problem dropping five or six hundred dollars on a better machine. Only after that, first great metal detecting experience.

Researching to find a great place to hunt

Where do you find treasure and relics? Now that you're metal detector is only forty percent of the equation. Research is the other sixty. Think about it, you can buy a thousand dollar machine anywhere. But it is very unlikely that you will find a gold ring, Barber silver dollar or, a confederate sword just anywhere.

So please put research over technology and budget. You should start by talking to old wise folks in your community. These folks have a ton of knowledge. So harness it. You can also use old pictures from historical books. Historical maps are also great for research. There is one map company called **Sanborn**. They created lots of historical maps for the nineteenth century. They were strictly for insurance companies. So they could be aware of fire risks when selling insurance. The correct name of these maps are **Sandborn fire maps**.

I have also had good luck with Google earth. You can log on to google earth then go back into time. I think to about nineteen- fifty. Find your house, and view what was there long ago. The Arial views are great. You can follow a river or creek. Then see where all of the wagon wheels made ruts in the water. These paths or crossings are called **fords.** They are great for water hunting.

You can also use old newspapers for stories about; where old circuses set up, or where some particular event happened. *Use your local historical society or the city's engineer's office.*

Decide What Type of Treasure You Want to Find

This all depends on what you have a passion for. Do you want to find civil war relics? If so that is great. Just check your local laws first. Yes there are laws that govern metal detecting in general. Remember they are subject to change from state to state.

Use common since and you will be fine. Don't dig in grave yards. Don't dig Indian burial grounds Just use a bit of common since and use common curtsey for your fellow man. One last bit of advice stay off federal property. Even if you are allowed to hunt there you are not allowed to dig without proper permits. Follow these suggestions and you will be fine.

There are places that are better to hunt certain things. You will be more likely to find coins and jewelry at school yards, public parks, and beaches. Any place, where a lot of people congregate is a good place to hunt. **People are constantly loosing items like coins and jewelry.**

Gold nuggets can easily be found in thousands of locations across the United States. Just research the published public mineral data bases. The treasure could be unlimited. So let's get out and start today.

Silver coins can be found any and everywhere. However these places are getting harder to search. Over the years these places have been pounded by detecorists. Remember a place is never entirely hunted out. They are just more difficult to hunt. There is plenty of unfound treasure everywhere. If you do not, find anything on your first trip. Go back with a different detector or another coil, you will be amazed about how much you can find after others have already searched. Soil conditions and conductivity is always changing in the ground.

On every hunt the parameters change a bit every day. This is due to corrosion and mineralization. My biggest tip on finding old silver coins is to find street and side walk tear outs. When the city tears out a street parking, lot or side walk, it is like opening very old time capsules. These capsules are great because there is no major modern trash on top of the targets to mask the signals.

Particularly the older parts of town are best. This is where the better treasure can be found. There is just so much treasure in the ground just waiting to be found. I am constantly surprised by the items that are lost by man in general.

Relics can be found everywhere. Plus they are easier to find than coins and jewelry. This is due to most relics being made from iron, lead copper, brass and bronze. This makes them much easier to hunt. That means you do not need a very expensive machine to relic hunt. Also you do not need as much discrimination. Any item left behind by people or an event, can be considered a relic.

Finding relics are very easy. You start searching until you find an item that able you to date your site. Square nails and coins are great for dating the site you're searching. The coins are a no brainer. Yet the square nails are a bit more difficult. It takes a bit more research. Square nails where made for hundreds of years. Through the years blacksmiths have hammered them out. During these years the technics changes thus this is how we date the site by using square nails.

After you have dated your site, search it in a more disciplined way. This is so that no part of the site gets left unsearched. *A grid search pattern works best for me*. After the first pass, you should have most of your trash removed from the first three to six inches. This is when the search becomes more exciting. You have now abled yourself to hear the sweet repeatable beeps of gold and silver. Crank up your discrimination and work the site thoroughly

How to dig targets ethically

In this chapter we are going to discuss digging your targets ethically. We all know that any dummy can take a digging tool from the shed. **Then head out and dig a hole**. This is not like professionals do it. Before starting to dig ask yourself these questions.

1. Can you dig with the intent to respect man and property?
2. Can you do it correctly? In a disciplined way, so that you are protecting Mother Nature and mankind? .

After you have pinpointed correctly, visualize your targets exact center location. This can be done by swinging your detector back and forth in a perfect straight line. Memorize the location that causes the beep. Next turn ninety degrees and do it again. This time mark the exact intersecting point of your swing lines. **This will be your targets exact location for digging**.

Now it is time to choose your digging tool. I use a Hori, Hori, knife (aka Japanese garden knife) or a Leche (digging tool designed specifically for metal detecting. I like using one of these tools for many reasons. They are very sharp and light weight. You can carry them easily in a sheath on your belt.

This allows one to be very quick and efficient, at cutting a plug. Remember the more plugs you cut the better chances you have at finding a good target. It also makes carrying your digger more comfortable. The biggest plus is you are less likely to lose your tool if it is located on your belt in a sheath.

How to cut a perfect plug

Cut a plug is defined *as cutting the grass in a circle pattern no more than six inches in diameter.* Be sure that all of the roots are cut. So that when you pick up the plug all of the soil is still intact with the roots. This is so that when it is replaced the grass does not die. Therefore you will leave no evidence of your dig.

Now that you have dug a hole, that can be easily backfilled. Next you will scoop all of the dirt from your hole until you have located your target. As you are scooping your dirt from the hole place the dirt on a cloth or towel. This makes it easier to poor back into the hole. If this step is skipped you will lose some of your backfill dirt into the surrounding grass. This will cause the backfilled hole to leave a divot or rut. That is not good; we want to leave the land scape just as it was before we dug.

You can use an electric pin-pointer to make finding your target much easier. After you have found the item that you were looking for. Check it again. Always double check your holes because change and jewelry are lost in lots. After you have back filled your hole, compact the dirt with the end of your digging tool. Then replace your plug. The last step is to stand up and stomp fiercely.

Please take the time to fill all of your holes properly. Leaving holes and trash for others is unethical and gives our hobby a bad reputation. It is also mandatory to take all of your trash with you. Never leave holes or trash for others to get hurt on.

By removing all of the trash from your site, you have made it possible for the good targets to be found by your metal detector. The trash has a big impact on hiding your targets. This is called **target masking**. Basically the trash can cover up the sounds of good targets thus making them harder for your metal detector to detect. **Remember everyone is always watching while you are detecting**. *You may not know it but they are.*

They are watching for many reasons. Some are better than others. They may only be nosey and inquisitive. They may be **interested** in your finds. Please believe they are also, waiting for you to mess up. So they can run you off, and call the police to say you are tearing up the park. **Just remember that someone is constantly watching you.**

Pin-Pointers

A pin-pointer is nothing more than a separate metal detecting device. That is used for finding your target inside your hole that you have dug. The reason we use a separate device is because our main detector does not fit into the hole that we have cut. The only problem I have ever encountered with a pi-pointer is that they do not have discrimination. They have no ability to distinguish a nail from a gold ring, unless it is a pin-pointer that is wired directly to the mainframe of your detector.

What does that physically mean? It means you need to check your holes twice. Especially if you find a nail because, there may be a ring in the same hole as your nail. Check again before you back fill your hole. Another problem encountered by many detectorist is they lose their pin-pointer easily. You will lose one in your lifetime especially if you dirt-fish somewhat often. Most seasoned detectorist use a lanyard to connect their pin-pointer to their belt.

Always keep fresh batteries in your pin-pointer. Low batteries will cause your detector to give erratic or false signals. Most pin-pointers are not water resistant either. So be careful when laying them in the grass or dropping them into a puddle.

Another Secret Tip

Although I do not like Garrett detectors, they do make the best pin-pointer on the market. They are quite costly about $127.00 at the time of writing this guide in 2014. Do not waste time or money on any other pin-pointer. Especially: The Minelab pin-pointer. It does detect one inch deeper, but isn't worth the exaggerated expense. Also it is not as rugged as the Garratt.

Here is another pin-pointer that we will touched on. It will cost about three hundred dollars for this technology. You can add a switch box feature to your main unit. This will allow you to use your main unit's discrimination modes while pin-pointing inside the plugs that you have cut. Sunpro makes the best switch boxes and parasite pin-pointer.

Water Hunting

Water hunting can be very rewarding. For obvious reasons, there aren't many people doing it. They are not doing it for many reasons. One of the reasons is that it can be extremely dangerous. I'm not going to go into much detail on this subject, because it is an advanced subject. If you are reading this it is because you just started detecting or you plan on starting.

I do want to share some tips; *Just in case you are interested in water hunting 101.*

1. Always wear foot protection some sort of shoe or boot. Preferably with a cleat. It can get very slippery when the algae grow on the rocks. You can make cleats out of many items like angle iron or nails or screws.

2. If you are using boots. Tuck your strings into your socks the current will keep causing then to untie. Please remember that there is more glass and sharp rocks in the water than treasure.
3. Never water hunt without tying your detector to yourself. For the currents will defiantly swipe it.
4. If you are not wearing a wet suit make sure to use sun screen. Everyone knows how the sun reflects off of water right into your face. This makes for bad sunburn.
5. Always take some sort of inner tube for use as a life preserver. Also makes for a good floating work station. Remember to tie it to yourself.
6. Use a water detector for water hunting. Under no circumstances ever take your land unit into the water. Remember different detector different job.
7. Try to take someone with you. If not possible leave a message with others about where you plan to go and when you plan on returning. This is for obvious reasons.
8. Last but not least. This is the most important to me because it almost killed me! **Do not use waist high rubber boots or bibs!** *They will fill with water and drown you.*

Needs Verses What I Want

When I first started metal detecting, I had to have the ace guru computer filled N.A.S.A. built machine. Wow! I was so wrong. Metal detecting is a numbers game period. The more plugs you cut. The more treasure you find. As a matter of fact the more complicated your detector is the less you will find. That is due to you believing that your two thousand dollar detector does not lie.

Believe me they do lie. There is no metal detector known to mankind that truly knows what is in the earth under your feet. Sure it can take a very educated guess. That is about it. That is all great in theory. But the gods honest truth is you will find more with a cheap box store metal detector than you will with the N.A.S.A. built machine.

Now you think I'm crazy or lying. An analog detector is a beep and dig machine. Literally when it beeps you dig. No fancy algorithms or nano micro processing technology. It does not have all the bells and whistles that a digital detector has. So if you follow that basic advice alone. You will single handedly dig more targets by yourself. Than two college grads with new N.A.S.A. built machines.

Digital VDI Liquid Crystal Display

Some metal detectors have a digital display. This display is usually a numerical reading. The reading indicates the scale that the metal detectors think that it is detecting. This is no more than an educated guess, of the metals conductivity that it has located.

However there are just too many mathematical parameters, laws of physics and chemistry, in play on each and every guess. On my detector there are four little boxes they are located on my l.c.d. screen. If one box is lit it means there is a twenty five percent chance that it knows the target. If two boxes are lit it thinks that there is a fifty present chance that it knows what it has detected. And so on. These boxes are called confidence indicators.

That is why I truly like my Whites MXT pro. It doesn't claim to be one hundred percent right all of the time. It does claim to be a fraction of a chance right sometimes.

So in closing on this topic I will say this, "Many machines are no better than the next". It is about learning the machine that you have now. Become one with your machine. You can always learn new technology and upgrade latter. Believe when I say that you will not feel nearly as bad digging up a pop can when your detector claims gold ring. As you would if, your two thousand dollar machine did the same.

Just get out and get started dirt-fishing. Don't worry about watching an l.c.d. computer screen. Buy the one you can afford now. Enjoy the weather and scenery. Get some exercise with your family and friends. Play the numbers game and you will find something special. *Welcome to the dirt-fishing community.*

Audio headphones

Audio headphones are important. They are necessary if you want to find the deepest smallest targets. Although it is possible to hunt without them, I recommend using them when you can. Make sure they are comfortable and fit well. The noise cancelling type is best. Please be careful about your surroundings when wearing them. It can get quite dangerous when hunting in the brush around rattle snakes and not being able to hear any sounds meant to warn you. Also be careful when metal detecting near streets or traffic. Please just use a bit of common since. It will take you far in this hobby.

Pick a pair of headphones that have a built in volume control. There are a lot of metal detectors with no volume control. So this feature on your headphones will help make them more comfortable. You will find that your cord to your headphones will be quite cumbersome. It will get in your way when digging. It will also break a lot and develop a short circuit. You will find that cordless headphones will seem like a valuable option. So I will provide how to do it in a bonus chapter at the end of this book.

Metal Detecting Salt Water Beaches

Metal detecting salt water beaches is much different than normal land hunting conditions. The difference is that it is quite difficult to hunt. So to be successful you will need different tools. Such as: A different and more expensive detector. Not only your machine is different but your digging tools are different also. You will need a sand scoop. The scoop works quiet well, but there will be some areas that have too many rocks or gravel mixed with the sand.

Please do not waste time and money on an aluminum sand scoop. The stainless Steele built ones are much better. While hunting beaches you will have to dig every target. You must search with little to no discrimination. The conductivity of the saltwater with the sand can be high. Therefore you will need a machine made for those conditions.

The P.I. machine is best for these conditions; the pulse induction technology is the best for working in very conductive situations. A whites surf master or a Garrett Infinium is a good choice. There are a couple of machines that are V.L.F. (very low frequency) technology that are capable of searching salt water beaches. **That machine is not the ace line by Garrett or the AT pro.** The AT pro is a great machine for hunting fresh water. The v.l.f. machines that are capable are top of the line machines like the MXT pro, Minelab Excailiber. The ATPro is able to water search water up to ten feet. Do not take your ATPro diving at 200 ft. Remember the deeper you take your machine;

the more atmospheric pressure is put on it. Just because it is water proof means nothing about dive proof. At two hundred feet your ATPro might implode, but it is more likely to just squeeze water into it and fry the circuitry.

You're Detectors Worst Enemies

The metal detector has quite a few enemies. Some are worse than others. Water can be your number one pain in the butt. It will ruin any detector if it gets into the circuitry. Water can also react with salt and cause conditions that drive normal detectors crazy.

Mineralization is another enemy to your detector. It causes detectors to have false signals and go crazy. Black sand, salt, iron and copper oxide are just a few minerals that will drive your detector insane. If your detector is not made for hunting these conditions it will make for a very bad experience. Then you may not want to detect anymore.

I know that all trash is bothersome an annoying. But there is no modern trash that is known to man to be more bothersome than aluminum soda and beer cans. Maybe I'm being a bit biased. Though you can ask any one that dirt-fish's," what he hates most?" They then will reply," aluminum cans and tabs". So it is important to choose your hunting sites carefully.

Recipe for a Successful Hunt

Now we are going to put the steps need to have a great hunt. First we will do all of our research about the site we, plan to hunt. Gather as much info about the site as possible. We should talk to older wise folks that have lived in the area from childhood.

Only after doing extensive research we gather all of our equipment. Detector, digging tools, first aid kit, finds pouch and belt. Do not wear steel toe boots. They will cause our detector to give false signals. Remember to check all of your local laws during research faze of preparation. This can turn into a night mare if this step is skipped. Local game wardens do not like," I did not know excuses "!

Wow! It's finally time to dirt-fish. It has been a long time coming. Take a good look at the site and study the layout. Use common since on where to begin your search. Look for old trees with exposed roots. This is a good indication that the landscaping has not been backfilled in a hundred or so years. Plus in the old days wise folks buried there money in mason jars. This was because they did not trust banks.

During the civil war people buried their valuables.
This was also common practice during the depression.

You can now hunt. Keep your metal detector low and swing slow. You need to swing your detector with, great patients making sure that your sweeps overlap each other. This is so that you do not miss targets. Search in any random pattern until you locate a good target. After that you will search in a spiral pattern out from your first find.

Remember to make certain that you ground balance your machine from time to time. The ground condition may change as you are moving around the site. **Fill any holes, and take all trash with you**. If you follow this recipe provided for you, chances are you will have a great hunt.

Not all trash is as seems

Not all trash is trash. There will be some treasure mixed in with your trash. So familiarize yourself with items that may look like trash but very valuable and sought after. Here are some examples; Oil cans from the early nineteen hundreds, Porcelain license plates, preferably three or four digit collectable ones. Old axe heads, parts from farm tools and old cars.

Keep all lead, brass, copper, and aluminum for recycling. All lead bullets from the civil and Indian wars are museum worthy. They can also be very valuable depending upon what battle they were fired. Old wagon wheels, any colored glass, depression era or earlier. Some, but not all shotgun shells are collectable. Kids toys like, hot wheels, frontier memorabilia, pocket knives, guns and musket parts are treasure. So with that said don't just trough away all trash thinking its trash. **Do some research**? *Some of your trash could truly be treasure.*

Conclusion

Now, that we have had all of this fun sharing information. You might think that I like to talk a lot. But let's put this all into perspective. Metal detecting can be very fun and rewarding. Iit is like anything in life. You will only get out of it. What you are willing to put into it. You may never find lost pirate treasure, but you will find treasure. And if you are willing to look hard enough, you just might find yourself like I did. I found that I liked it so much I needed to share it with others. I hope this E-book has been helpful. Remember if you every have any questions on this subject. I can always be found at www.metaldetectingfordummies.com www.urbantresureatwholesale.com.

Happy dirt-fishing

Bonus chapter

Design cordless Headphones for your detector

This is a really cool hack that I've figured over the years. There is a certain need for this hack because the cord on most headphones is very brittle and, seems to get in the way when digging targets. The cord also gets cut accidentally by your digging tool. It can also get caught under your foot when your kneeling cutting plugs. This intern causes the cord to be ripped out of the ear piece.

To make this project a success, you will need these items listed below.

1. An inexpensive set of cordless headphones made for a home audio system. The cost nearly fifteen dollars and come complete with the transmitter.
2. A quarter inch audio plug(jack)
3. About two feet of audio wire. (Old rca cables work great).
4. Soldering iron and solder.
5. Black tape or shrink wrap.
6. Hot glue gun.

Before we begin we must make sure that the cordless headphone transmitter is a dc unit and not an ac unit. Three to nine volts will work fine. You must disassemble the transmitter and remove the circuit board and battery box. Get rid of the huge plastic factory housing. Then you can decide whether you want to install the transmitter circuitry into your main unit or create separate parasite housing for it (my first choice). If you have decided to use a separate parasite housing, then you must figure where you're going to mount it. I always mount mine under the arm support. (See photo) like shown on photo on the bottom of this page. This is where you will use the hot glue gun to mount your device to the detector.

After mounting your fm transmitter to your detector, hook up the correct power supply that is needed to drive your unit. Now it's time to solider your wires to the input of the transmitter. Next solider the male audio jack to the other end of your wires. Use the black electrical tape to insulate the wires where they have been soldiered. You could also use the shrink tube but not necessary.

If you have chosen to install the transmitter inside the mainframe of your unit, that is neat. But be sure the circuitry is tucked neatly out of the way of your detectors components. You will solider the plug end of the transmitter to the speaker contacts instead of the quarter inch jack. Button your unit back up. Now you're in business.

Put the headphones on your head. Turn on your detector and transmitter. Then press the scan button on your fm cordless headphones. Your headphones will then find the channel or frequency that your detector is broadcasting on. Now you have cordless headphones on your one hundred dollar entry level machine.

This option is only available on top end two thousand dollar models. I hope this chapter somewhat instructs you on how to do it. I did not go into full detail, because I wrote this guide thinking that you already have a basic knowledge of electrical projects and soldiering. I have a full instructional step by step video on you tube if you need more information. Here is the link. http/www.roseberrywilliam.com

Author's Notes

First and foremost **I would like to thank you** for supporting my endeavor, by purchasing this book. **To my family, I want to say**, "Sorry for taking quality family time away from ya'll", during the research and creative process. **Anyone seeking further information on metal detecting**, I recommend the Lebanon county library. *They have a wealth of information available on the subject....*

Made in the USA
Monee, IL
04 November 2026

67385560R00026